DATE DUE

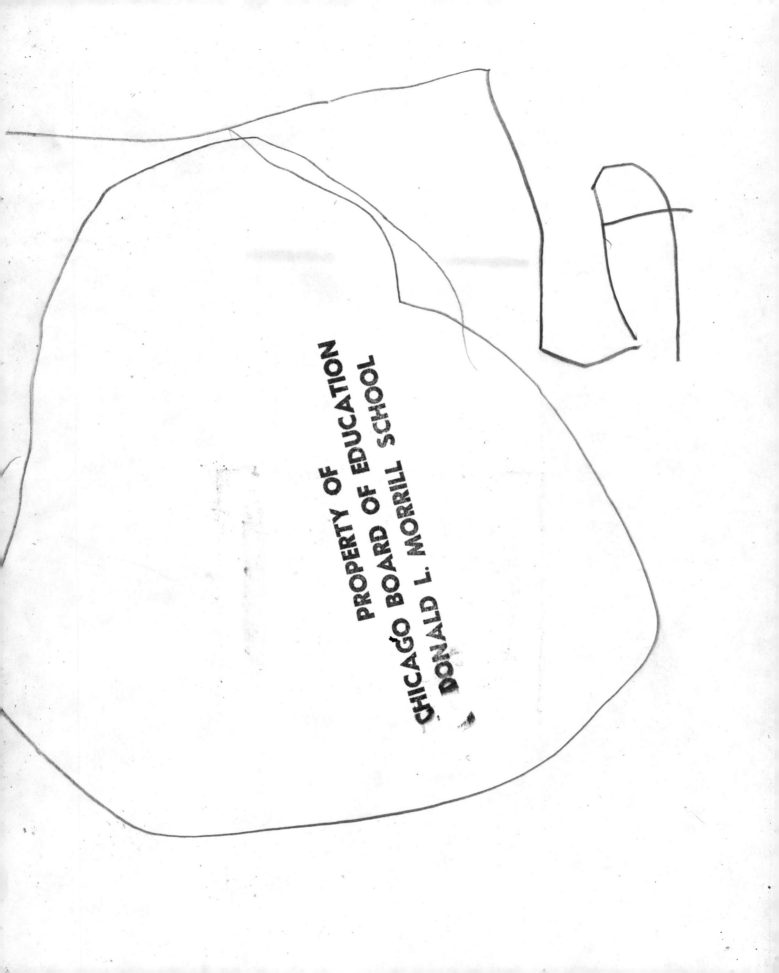

GEOTHERMAL AND BIO-ENERGY

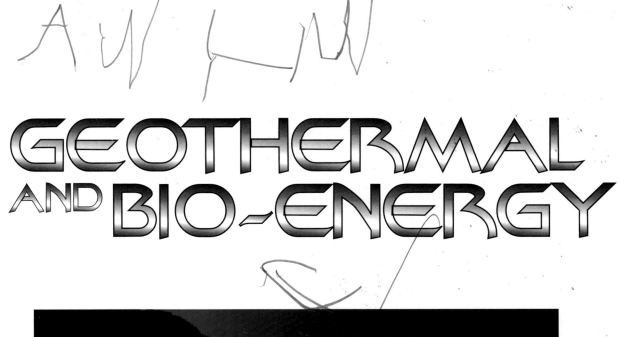

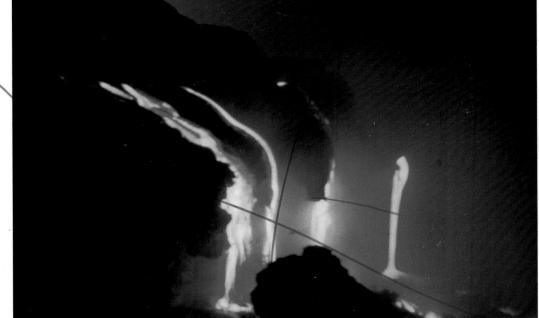

IAN GRAHAM

FSVP
RAINTREE
STECK-VAUGHN
PUBLISHERS
A Steck-Vaughn Company

Austin, Texas

ENERGY FOREVER?

Geothermal and Bio-energy

OTHER TITLES IN THE SERIES

Solar Power · Water Power · Wind Power
Nuclear Power · Geothermal and Bio-Energy

Published by Raintree Steck-Vaughn Publishers,
an imprint of Steck-Vaughn Company

Library of Congress Cataloging-in-Publication Data
Graham, Ian.
Geothermal and Bio-energy / Ian Graham.
    p.   cm.—(Energy Forever)
Includes bibliographical references and index.
Summary: Defines geothermal and bio energies,
and explains their advantages and disadvantages.
ISBN 0-8172-5367-X
    1. Geothermal resources—Juvenile literature.
    2. Biomass energy—Juvenile literature.
    [1. Geothermal resources. 2. Biomass energy.]
    I. Title. II. Series.
    GB1199.5.G7   1999
    333.8'8—dc21   98-38726

Printed in Italy. Bound in the United States.
1 2 3 4 5 6 7 8 9 0 03 02 01 00 99

Picture Acknowledgments

AEA Technology, Harwell: 11, 16, 17 top, 17 bottom, 23 top, 25,
27 right, 29. Ecoscene: 1 and 12 (Nichol), 4 (Gryniewicz), 7
(Ayres), 9 (Alan Brown), 10 (Sally Morgan), 10–11 (John Corbett),
27 left (Ford), 28 left (Platt), 28 right (Gryniewicz),31 (Jim
Winkley), 33 (Jim Winkley), 35 (Jim Winkley), 36 (Kevin King), 40
(Joel Creed), 41 (Sally Morgan), 42 (Moore), James Davis Travel
Photography: cover, 5, 19, 30 left, 30 right. Eye Ubiquitous: 14
(David Cumming), 24 (Dean Bennett), 34 (L. Johnstone). Olë Steen
Hansen: 6, 32. Mary Evans Picture Library: 15. Oxford Scientific
Films: 18 (Richard Packwood), 43 (T. Middleton). U.S. Department
of Energy: 20, 23 bottom, 38, 39 left, 39 right.

CONTENTS

TWO DIFFERENT KINDS OF ENERGY

A piece of camelthorn bursts into flames when it touches hot rock in the Timanfaya National Park on the volcanic island of Lanzarote in Spain's Canary Islands. In such places, magma (molten rock) is very close to the earth's surface.

What is geothermal energy?

Geothermal energy is heat deep underground that can be used to produce electricity. This electricity is called geothermal power.

A volume of rock about the size of an average house that lies below the earth's surface and is several hundred degrees hotter than the surface rock contains as much energy as the whole world uses in one year. But to be useful, that heat energy has to be converted into a form, such as electricity, that can be transported to where it is needed. Geothermal power plants use geothermal energy to make electrical power.

Geothermal power around the world

Currently, about 25 countries use geothermal power. The world's largest user is the United States, with a production capacity of about 3,000 megawatts. The next-largest user is the Philippines, which produces 1,200 megawatts, about one-fifth of its total electricity production.

Other users of geothermal power include New Zealand, Russia, Mexico, Italy, Japan, Indonesia, Costa Rica, El Salvador, and Turkey. The total output of all the world's geothermal power plants is about 7,000 megawatts. This represents 0.15 percent of world electricity production. Most electricity is produced by hydroelectric power plants or by the burning of fossil fuels—coal, oil, and natural gas—in power plants.

At Svartsengi geothermal power plant in Iceland, steam rises from the turbine-room chimneys, and people bathe in the "Blue Lagoon'" of warm water beside the plant. Water heated to steam by hot rocks underground turns turbines that are linked to generators, which produce electricity. The steam condenses to water, which empties into the lagoon.

What is bio-energy?

Bio-energy is power produced from biomass. Biomass is short for biological mass, which is another name for plants and animals.

The world's plants are a rich source of energy. And of course, by producing seeds, which creates new plants, they are constantly renewing themselves. Enough plants grow every year to meet the world's energy needs eight times over. But at present, we use only about 7 percent of the annual plant production.

Green plants include ferns, conifers, and all flowering plants. They all make their own food, using sunlight, water from the soil, and carbon dioxide from the air.

Biomass is a vital raw material in the Third World. Plant material can be burned for light, warmth, and cooking, woven to make clothing and ropes, or shaped into weapons and tool handles. It is also an essential building material.

Green leaves make sugars and starch by using light energy to combine carbon dioxide from the atmosphere with water from the plant's roots. Oxygen is released as a by-product of the chemical reactions.

The energy within the sugars and starch is released as the plant is eaten, rots, is burned, or is chemically digested.

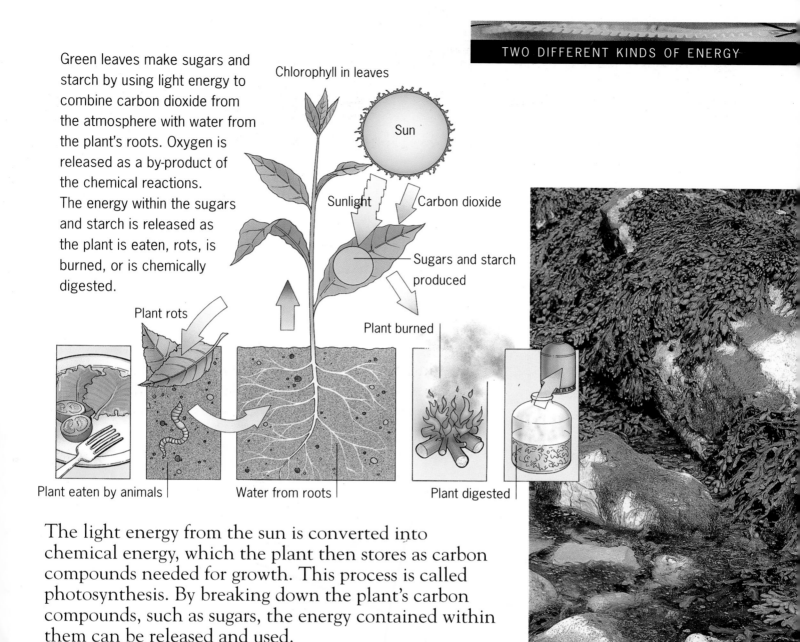

Chlorophyll in leaves

Sun

Sunlight

Carbon dioxide

Sugars and starch produced

Plant rots

Plant burned

Plant eaten by animals

Water from roots

Plant digested

The light energy from the sun is converted into chemical energy, which the plant then stores as carbon compounds needed for growth. This process is called photosynthesis. By breaking down the plant's carbon compounds, such as sugars, the energy contained within them can be released and used.

Electricity from biomass

Energy from biomass can be converted into other forms of energy, particularly heat and electricity. About 4 percent of the electricity produced in the United States comes from biomass. That is about the same amount of electricity as is produced by nuclear power plants or hydroelectricity. Sweden and Canada each meet about 8 percent of their national power needs from biomass.

The sea provides a rich harvest of biomass in the form of seaweed. But while seaweed is often eaten or used as a fertilizer to feed crops, it has so far been little used in bio-power plants for producing electricity.

GEOTHERMAL AND BIO-POWER

Where does geothermal power come from?

The word "geothermal" comes from two ancient Greek words—geo and therme—meaning, "Earth heat." The center of the earth is made of a solid ball of metal (iron and nickel) at a temperature of about 7,600 ° F (4,200° C). It is so hot that it melts the-surrounding rock.

Most of the energy that heats the earth's core to such a high temperature comes from nuclear reactions taking place there. Fortunately for us, the surface of the earth is much cooler than the core. We live on the cool rocky crust that floats on top of the superhot core.

The earth's core is divided into a solid metal inner core surrounded by a liquid metal outer core. This is surrounded by the mantle, which moves and flows like warm toffee. Finally, outside the mantle is the thin solid crust. The crust is thickest beneath the continents and thinnest beneath the sea.

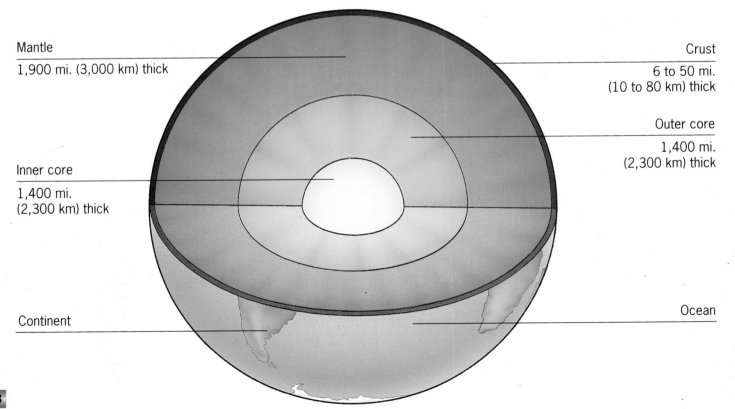

Mantle
1,900 mi. (3,000 km) thick

Crust
6 to 50 mi.
(10 to 80 km) thick

Outer core
1,400 mi.
(2,300 km) thick

Inner core
1,400 mi.
(2,300 km) thick

Continent

Ocean

Geothermal heating

As heat flows away from the core out to the earth's surface, its temperature drops as it travels. When the heat reaches the surface, it is carried away by the atmosphere and finally radiated away into space. Normally, we do not notice geothermal heating, because the sun has a greater heating effect on the surface than the earth's core does.

Tenerife, one of Spain's Canary Islands, is made of volcanic ash and lava. The snowy peak in the background is Mount Teidi. Rock like this, formed by heat, is called igneous rock.

Lichens cover the trunk of a giant Redwood tree in California. Lichens are found farther north and farther south than other plants and at greater altitudes. They can grow on mountains at up to 20,000 ft. (6,000 m). The local people use lichens as fuel.

Where do biofuels come from?

Biofuels are fuels made from biomass. They are burned or processed with chemicals or microorganisms to release the energy stored inside them. Biomass includes all plant material—forest trees (both the woody trunks and the leaves), shrubs, grasses, peat from bogs, seaweed from the oceans, and mosses and lichens from polar tundra. It even includes animal dung, which is often rich in undigested vegetation.

Nature's fuels

Biofuels grow all around us. The coldest, hottest, and driest places on Earth—the frozen poles and the searing, arid deserts—support very little plant life, but the rest of the earth teems with a rich and varied covering of plants. Forests stretch all the way from Canada's Pacific coast to the Atlantic and from Scandinavia eastward across northern Europe and Russia. The Amazon rain forest covers huge areas of South America. Seaweed flourishes in the fertile coastal shallows around the continents. Lush jungles thrive in tropical regions with high rainfall. Trees, shrubs, grasses, and other plants cover much of the countryside in temperate climates.

Forest and grassland stretch into the distance across Lapland, Sweden. Even in these cold northern climates, vegetation grows everywhere. All this biomass could be used as biofuel by burning it or by processing it chemically to make gas or oil.

Pine trees are harvested in Scotland as biofuel. The bark, waste wood chippings, and sawdust can be burned as a fuel to make electricity.

11

How does geothermal energy affect us?

The physical world around us seems constant and unchanging. In fact, it does change but only extremely slowly. The changes in the natural world are caused by geothermal energy. It drives the continental drift that produced the mountain ranges, steep rift valleys, and ocean basins. Normally, the effects are very gradual, taking millions of years to make a real difference. But geothermal energy can also be sudden and violent, as when magma (molten rock) near the earth's surface breaks through the crust to produce a volcano.

Volcanoes

When a volcano erupts, rock, dust, and ash are flung into the air and rivers of lava—magma that reaches the surface—flow down its sides. Clouds of searing hot dust and gas, called pyroclastic flows, can hurtle down the volcano's side at up to 300 mph (500 km/h), incinerating everything in their path.

FACTFILE

There are about 850 active volcanoes around the world. Most of them are located around the edges of the Pacific Ocean in a region called the Ring of Fire.

Rivers of red-hot lava light up the night in Hawaii. The Hawaiian Islands were formed from volcanoes that grew up from the ocean floor and rose above the sea. Nowadays, Mauna Loa and Kilauea are the only active volcanoes in Hawaii.

If the top of the volcano is capped with snow, the snow, melted by the hot magma, can mix with ash and wash down the mountainside as lethal mudflows that can bury whole towns. Volcanic eruptions unleash enormous amounts of energy, but they are such violent events and so infrequent that it is not possible to harness their power.

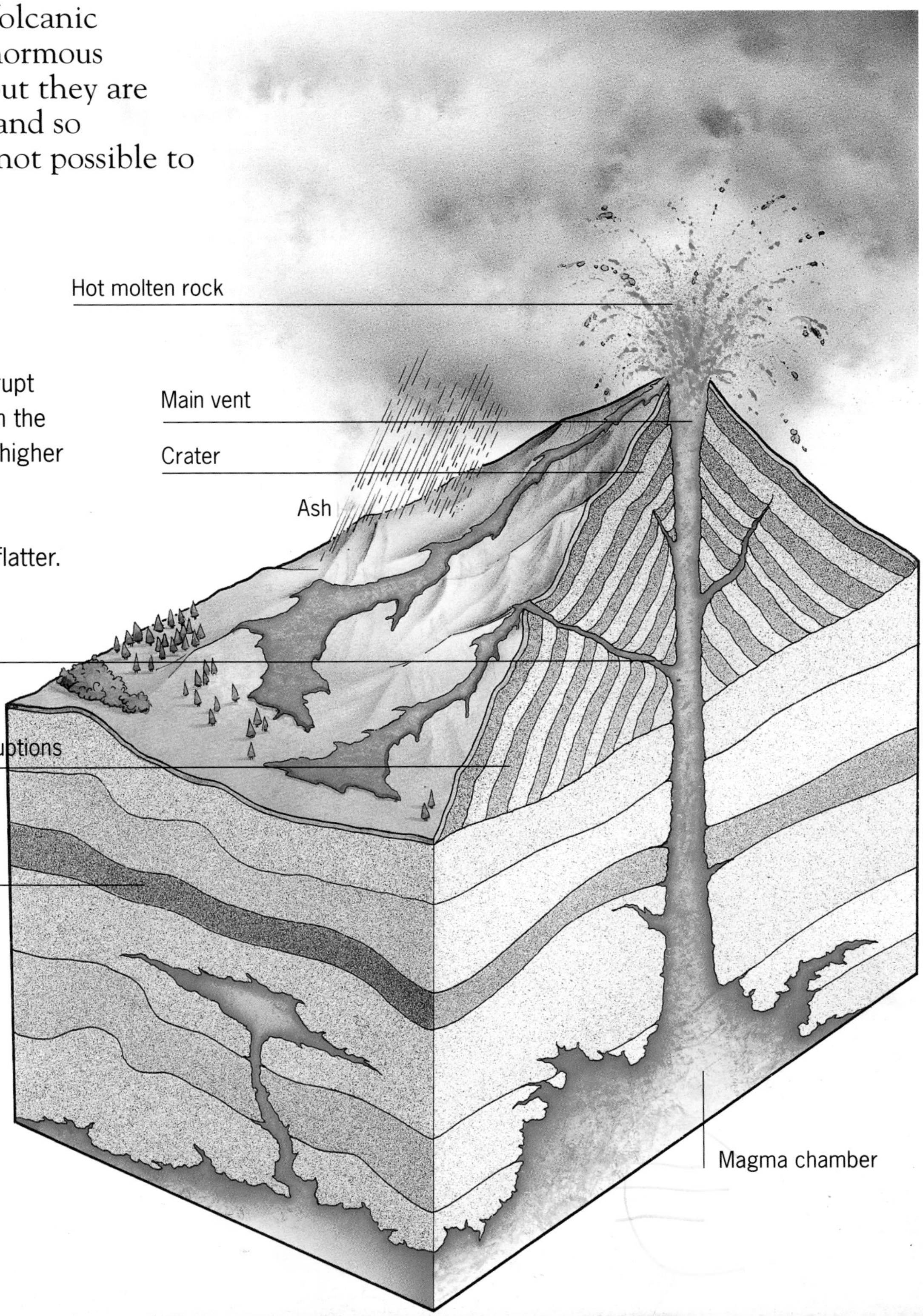

Cone-shaped volcanoes erupt regularly. Lava pours down the sides and builds the cone higher and higher. Other types of volcanoes, such as shield volcanoes, are lower and flatter.

Hot molten rock

Main vent

Crater

Ash

Side vent

Layers of lava from earlier eruptions

Rock layers

Magma chamber

GEOTHERMAL AND BIO-ENERGY IN HISTORY

Harnessing bio-energy

People have warmed themselves and cooked their food over wood fires for thousands of years. In some developing countries, wood fires are still the only means of heating and cooking. Burning is also important in developed countries, where household trash is burned in incinerators to make electricity. Biogas—gas from naturally decaying biomass—has also been used as a fuel. As long ago as 1857, methane gas from decaying biomass was used as a fuel in a leper colony near Bombay, India.

A family in Rajasthan, northwest India, huddles around a wood fire for warmth. Wood is still the most important fuel for millions of people today.

Geothermal power in history

The Romans built 57 hot spring baths across their ancient empire, from north Africa to northern England and from Spain to Turkey. They believed the hot, mineral-rich waters were good for their health.

However, in 1903 the world's first geothermal power plant was built at Larderello, in Italy, on the site of ancient baths. Electricity production began the following year. Today, Larderello power plant generates 390 megawatts, which is enough to power a small village.

Citizens of the German town of Leuk take the hot spring waters in the eighteenth century. Fully clothed, they stayed in the water for hours, passing the time by reading books or newspapers or playing chess on floating boards. Others visited the baths to breathe in the vapors given off by the water.

15

What is peat?

Peat is made from partly decayed plants in wetlands called bogs. Mosses growing on the surface die, sink, and are buried and squeezed by new plants growing above them. With little oxygen present, bacteria cannot completely break down the dead plants. The remains, which are rich in carbon, build up into peat.

Left for millions of years, the peat would turn into coal. Peat bogs can be very deep and are found all over the world. They cover one-third of Finland and one-tenth of the Irish Republic. Canada and Russia are also rich in peat.

Peat briquettes and milled peat are among the products made by the Irish Republic's Bord Na Mona Peat Development Authority. Most of the milled peat is used in peat-burning power plants that supply the country's electricity. Briquettes are used for heating in homes, offices, and factories.

Hand-cut turfs of peat are being examined by a quality controller. Peat has many uses besides being a fuel. It has been used as bedding, building material, and fertilizer and, recently, because of its absorbent powers, as a treatment for oil spills.

Peat as fuel

Burning peat as a fuel for the home dates back at least to Roman times. It is still widely used, particularly in areas where other fuels are in short supply. Today, most of Finland's inland cities are heated by peat-fired power plants. In the Irish Republic peat provides the energy for one-fifth of the country's electricity. Russia opened its first peat-burning power plant as long ago as 1914. But the United States did not begin using peat in a big way until 1990, when the country's first peat-fired electricity-generating plant was opened in Maine. Today this contributes 22.8 megawatt-hours.

F A C T F I L E

Biological activity is so slow in the wet, acid conditions of peat bogs that animals that fell into them hundreds, or even thousands, of years ago can be found today almost perfectly preserved. Hundreds of dead human bodies have also been found in bogs all over Europe, some of them more than 2,000 years old. Many victims were killed deliberately, perhaps as offerings to their gods.

Peat is dug out of the ground by machine, cut into blocks and left to dry. Once upon a time it had to be dug out by hand. The dried blocks can be burned as fuel.

GEOTHERMAL AND BIO-ENERGY TECHNOLOGY

Wells

Hot rocks lying within a mile or so of the surface can be reached by drilling. Any underground water that passes through cracks in the rock will be heated by the rock, sometimes to 300° F (150° C) or more. The hot water can be extracted from the well and the heat energy used to make electricity.

Water and mud heated by hot rocks underground collect in pools near Lake Myvatn in Iceland.

Water is pumped under great pressure from a geothermal power plant into a region of hot rock. The pressure forces the water through small cracks in the rock until it reaches the extraction well and rises to the surface.

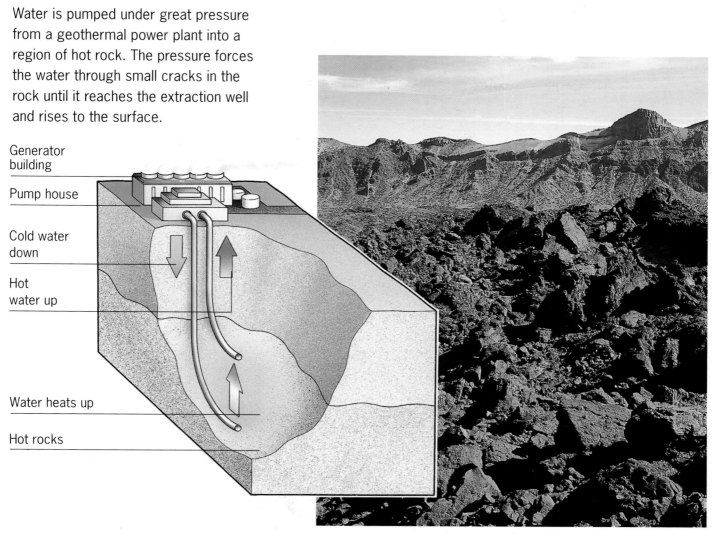

Generator building

Pump house

Cold water down

Hot water up

Water heats up

Hot rocks

Artificial wells

If the hot rocks underground are dry, an artificial hot well can be created. Two holes are drilled some distance apart. Cold water is pumped down one hole and is extracted as hot water or steam from the other. Underground the water moves between the two holes, collecting heat on the way.

If necessary, the rock can be shattered with explosives to allow the water to pass through. Sometimes the water is hot enough to turn to steam. When the steam reaches the surface, the heat is extracted from it, and it is condensed back into water. The water is then sent back underground again.

The moonlike lava fields on the Spanish island of Tenerife show that a volcano once erupted there. But by now the magma may have retreated too far underground to make it worthwhile to try to develop geothermal power.

The Geysers Geothermal Power Plant, California, was developed as a commercial supplier of electricity following the success of geothermal energy research at Los Alamos.

Los Alamos National Laboratory

In 1986, two experimental wells 2.5 mi. (4 km) deep were drilled at Los Alamos National Laboratory, New Mexico. Cold water was injected into one well. At the bottom, the water filled a natural cavity in the rock, where it was heated by geothermal energy.

The pressure in the cavity pushed the water up and out through the second well, the extraction, or production well. To stop the cold water from running straight through, the bottom of the production well was positioned higher than the bottom of the injection well. Thus the water could not reach the production well before it had been thoroughly heated in the cavity.

Megawatts from heat

When the water emerged at the surface again, its temperature had been raised to 375° F (190° C). The water was then passed through a device called a heat exchanger, to extract its heat energy. The heat was converted into electricity and the cooled water was pumped underground again. The heat energy recovered was converted into 4 megawatts of electricity.

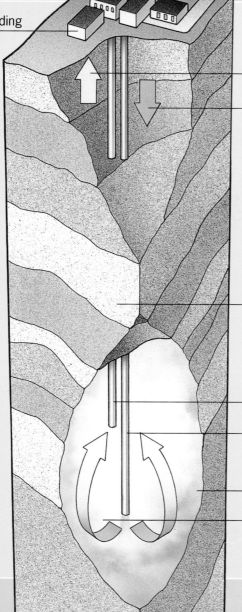

Generator building

Steam up

Water down

The temperature in some underground cavities can reach 575° F (300° C). The Los Alamos experiment showed that, using water as a "transport medium" to collect the energy and bring it to the surface, an immense reserve of energy could be tapped to generate electricity.

Rock layers

Steam up pipe

Water down pipe

Natural cavity in rock

Water turns to steam

Burning and burying biomass

In countries where wood is plentiful, such as Canada, wood-burning stoves are used routinely for cooking, and homes are heated by wood-burning furnaces. In sawmills, waste wood is burned, and the heat is used to dry new lumber before it is cut. In pulp and paper mills, the waste heat is used to generate steam.

There is quite a lot of biomass in household waste, and this has traditionally been buried in landfill sites. But the amount of waste produced by developed countries has grown enormously. Finding new places to bury it all is becoming a problem.

Recycling and incineration

The amount of waste that has to be buried can be greatly reduced by recycling materials and burning the rest in incinerators. Incinerators work at high temperatures and extract much energy from the waste. The high temperature also neutralizes toxic gases given off, for example, by plastics in the waste. To ensure that the biomass burns efficiently, air is pumped through a bed of sand at the bottom of the incinerator to provide a constant and plentiful supply of oxygen.

FACTFILE

Great Britain's first large-scale waste incinerator designed for energy supply went into service in London in 1994. It consumes about 500,000 tons of solid waste each year and generates 30 megawatts of electricity—enough to supply more than 60,000 homes. First, the waste is sorted to remove dangerous or unwanted items and anything that can be recycled. The rest, mostly organic waste, paper, and plastic, is then burned to release energy. This fuel is called RDF (Refuse Derived Fuel).

Sawdust and other wood waste from a sawmill are poured into a combustion chamber, or boiler, and burned. The heat changes water to steam, and this drives a turbine and electricity generator. The steam is then condensed back into water, which is returned to the boiler. On its way, the steam is used to heat the sawmill building and to dry wood chips. These are compressed to make artificial logs for sale.

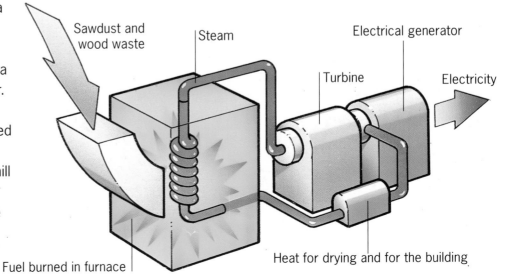

Sawdust and wood waste | Steam | Electrical generator | Turbine | Electricity

Fuel burned in furnace | Heat for drying and for the building

Above: Straw burns inside an incinerator to create heat. Straw is the dried stems of crop plants such as wheat, barley, and oats.

This power plant in California burns wood and farming wastes to make electricity. Heat from the burning waste changes water to steam, which drives turbogenerators to make electricity.

Bug power

Sometimes, organic material—biomass—is left to decay and be broken down by bacteria. In India and China, people often collect animal dung and put it into tanks. Bacteria break it down and give off gas, which is used for cooking, heating, and even driving generators.

The bacteria that do this job are anaerobic—that is, they grow best where there is little or no oxygen. They are responsible for the eerie glow, called "will o' the wisp," that sometimes appears over swamps. The gas from the bacteria breaking down vegetation in the swamp bubbles to the surface and occasionally catches fire, producing the glow.

Cattle draw a plow across a paddy in Bali, Indonesia. Animals kept for food and farming in large parts of Africa, Asia, and the Far East produce vast amounts of waste that can be dried and burned as a fuel.

Biogas power

A rural home could supply all its energy needs from the biogas produced from waste supplied by 800 pigs or 100 cattle. Human waste can be turned into fuel gas in the same way. Methane gas from sewage contributes 33 megawatts of electricity in Great Britain—about 2 percent of the country's total electricity needs. Methane from garbage adds another 80 megawatts, and burning trash another 130 megawatts.

Right: A technician checks a pressure meter at a methane production plant. The methane is produced by bacteria in a sewage works.

A methane digester is just a large tank. When slurry (animal waste) is poured into the tank, the bacteria change it into simpler sugars and acids and then into a gas, which can be burned as fuel.

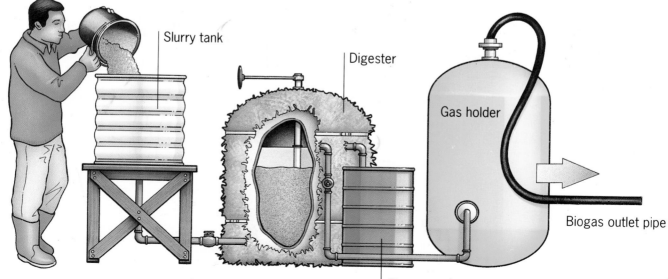

Slurry tank

Digester

Gas holder

Biogas outlet pipe

Slurry overflow tank

Environmental impact of bio-energy

Forests and peat bogs are disappearing all over the world as people take wood and peat for fuel. Peat bogs take thousands of years to form, so removing peat from them destroys them. Trees can be renewed by replanting them with fast-growing species, but instead of replanting, people often just look farther afield for wood.

There is a serious fuelwood crisis in many parts of the world, especially central and southern Africa, the west coast of South America, Nepal, and northern India. Stripping the land of trees can also destroy the soil by allowing it to be blown or washed away by wind and rain. Tree roots bind soil particles together.

Wood is the most popular fuel for cooking and heating in the Third World. As populations grow, the need for firewood also grows, but there is not always enough wood to meet this growing demand. This map shows how widespread firewood shortages have become.

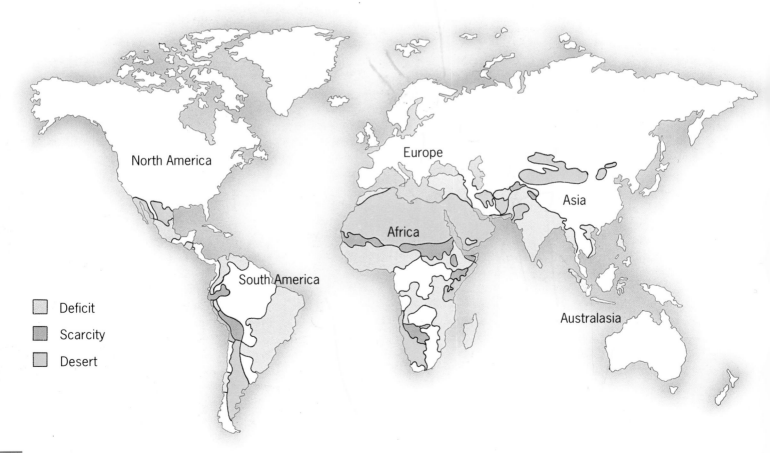

North America

Europe

Asia

Africa

South America

Australasia

Deficit

Scarcity

Desert

Global warming

Biomass contains carbon. Burning biomass releases the carbon into the atmosphere as carbon dioxide gas. Carbon dioxide is a "greenhouse gas." Greenhouse gases are thought by many scientists to cause global warming. However, burning biomass harms the environment less than burning fossil fuels such as coal or oil. It merely recycles today's carbon dioxide instead of releasing extra carbon dioxide that has been locked away for millions of years.

Below: Collecting firewood in Kerala, southern India. People in many parts of the Third World face the daily problem of finding enough wood to fuel their fires.

Below: Sacks of household garbage are piled together at a landfill site. The garbage will be covered with earth, and the gas from rotting biomass will be used as fuel.

Above: Natural hot water, supplied directly from geothermal sources or as waste water from a geothermal power plant, can heat a greenhouse.

Above right: Steam rises from the condensers and pipelines at Olkaria geothermal power station near Naivasha, Kenya.

Environmental impact of geothermal energy

Geothermal power plants are kinder to the environment than other power plants. They use a renewable energy source and produce little or no harmful gases or dangerous waste materials. Geothermal stations also release less nitrous oxide, which forms ozone gas. Ozone is very desirable high in the atmosphere, because it shields us from the harmful ultraviolet radiation in sunlight, but at ground level it is a pollutant.

Disadvantages of geothermal energy

Some people are concerned about taking large amounts of water from the ground. Water used by a geothermal power plant must be reinjected into the ground to keep the water table from falling. The water table is the natural level of water in the ground, and if this is allowed to fall, the land could shrink and sink.

Also, underground gases that are allowed to escape into the air can smell bad and can cause air pollution and excessive noise. When steam and gases are released from pressurized tanks before cleaning them, the noise has been compared to a jet aircraft taking off.

F A C T F I L E

A 30-megawatt geothermal power plant on the island of Hawaii reduces the island's oil imports by 500,000 barrels a year. Burning less oil further cuts Hawaii's carbon dioxide emissions by 220,000 tons. Oil is delivered by tanker, so reducing the need for oil also reduces the risk of oil spills at sea.

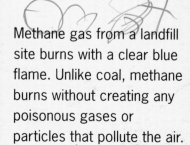

Methane gas from a landfill site burns with a clear blue flame. Unlike coal, methane burns without creating any poisonous gases or particles that pollute the air.

Above: Mud bubbles like boiling gray soup in Gullbringu on the Reykjanes Peninsula in Iceland. Hot mud pools are common in areas rich in volcanic activity.

When the Romans discovered natural hot springs in England in the first century A.D., they built baths near the site. The town that grew up around them was called *Aquae Sulis*. It later became known as Bath.

Hot springs

Hot springs are created where water heated by hot underground rocks or volcanic activity comes to the surface. They are common in Iceland, New Zealand, and other volcanic areas around the world such as Yellowstone National Park and Hot Springs National Park in the United States.

In some volcanic regions of the world, finely powdered rock particles mix with hot spring water and produce bubbling hot mud pools called mud pots or mud volcanoes.

Spring heating

In parts of Reykjavik, the capital of Iceland, buildings are heated by water piped from nearby hot springs. The hot springs in and around Baden-Baden in Germany were discovered and developed by the Romans and have been used for bathing and drinking for 2,000 years.

The Romans believed that geothermally heated water could relieve or cure common ailments. They not only built many baths, but also heated their bath buildings by piping through the hot vapor from the springs.

FACTFILE

In 1883, a party of workers building the Canadian Pacific Railroad saw what they thought was smoke rising from a hole in the ground. When they looked closer they found it was steam from natural hot springs. The springs drew visitors. A hotel was built for the them and the town of Banff grew up around it. It was the first geothermal resource to be developed in Canada, and it became Canada's first national park.

At Mammoth Springs in Yellowstone National Park in Wyoming, the gently flowing mineral-rich waters have shaped the rock into broad flat terraces up to 300 ft. (90 m) high. As the water flows over the rock, it leaves behind a deposit of travertine, a type of limestone. Brightly colored algae grow in some of the pools.

This geyser on the island of Lanzarote shoots out water and steam with tremendous force.

USES OF GEOTHERMAL AND BIO-ENERGY

Geysers

One of the most spectacular displays of geothermal energy is the geyser. Water trickles down a crack in the ground and meets rock so hot that the water at the bottom boils. The hot water pushes the water above it upward. When the water rises, there is less weight pressing down on the bottom of the water column.

With the pressure now reduced, the boiling water flashes into steam, and the sudden expansion of the steam blasts out the rest of the column. It shoots out of the ground as a tall, noisy jet of steam and water. The height of a geyser can range from a few feet to about 330 ft. (100 m), and the interval between eruptions ranges from a few seconds to several hours.

A world of geysers

Almost all the world's geysers are found in New Zealand, Iceland, the United States, and Russia. They are usually found in areas of volcanic activity. In Iceland, there are several dozen geysers within a mile or so of each other near Reykjavik, Iceland's capital city.

The word geyser comes from Geysir in Iceland, one of the world's most spectacular geysers. It throws up a column of water 200 ft. (60 m) high every 5 to 36 hours. Erupting much more frequently, the Strokkur geyser, also in Iceland, performs every three minutes, while the Velikan geyser, on the Kamchatka peninsula, Russia, erupts at intervals of three hours.

Water and steam spout from a geyser in the Norris Geyser Basin in Yellowstone National Park. The park has the greatest concentration of geysers in the world.

Old Faithful

In the state of Wyoming, is Yellowstone National Park, the oldest national park in the world. Most of its breathtaking landscape was laid down by volcanoes, which erupted there more than 50,000 years ago. Molten rock still lies so close to the surface that it has created thousands of spectacular hot springs, boiling mud pools, and geysers. The most famous of the park's geysers is Old Faithful. It is not the tallest geyser in the world but, as its name suggests, it is one of the most predictable.

Thousands of tourists visit Yellowstone National Park every year to watch Old Faithful in action. They are not disappointed, because the geyser never fails to perform.

Old Faithful lies dormant, but not for long. Deep underground, water is flowing down toward the molten magma that will flash the water to steam and blast it out of the ground along with thousands of gallons of water lying above it.

Old Faithful's vital statistics

Old Faithful erupts every 37 to 93 minutes, shooting up to as high as 170 ft. (52 m). The precise interval between eruptions depends on the strength of the previous eruption. There is a long pause after a big blast.

Each time Old Faithful blows its top, it blasts about 10,600 gal. (40,000 l) of water into the air. Just before the main eruption, it produces a series of small jets of water up to 23 ft. (7 m) high.

Geothermal power plants

Most of the world's geothermal power plants are in Italy, New Zealand, the United States, Mexico, Japan, and Russia. But many other countries are beginning to develop geothermal power. El Salvador was the first Central American country to do so, and in the Pacific the Philippines has become the world's second-largest geothermal power producer after the United States.

Power from heat

Sometimes underground steam can be used directly to drive the power plant's turbines and electricity generators, as at The Geysers in California. However, most geothermal power plants have to use underground water to do the job. If the water is hot enough, it can be changed into steam simply by lowering the air pressure over it, which gives the steam space to expand.

Pipes are laid under a road in Reykjavik, Iceland. The pipes will carry hot water from a nearby geothermal source. As the heat spreads out from the pipes into the surrounding ground, it will help keep the road surface free of ice.

Often, the water from underground is cooler than this. It is used to make a vapor by heating a second liquid that boils at a lower temperature than water. The vapor then drives the turbines. Once the steam or hot water has passed through the power plant, it is still hot enough to heat buildings or greenhouses.

When the underground water is between 212° F (100° C) and 350° F (175° C), a binary plant is used. Here, the water is used to evaporate a second liquid with a lower boiling point than water. Vapor from this second liquid drives the turbines. The vapor is then condensed back to liquid, which is used again.

F A C T F I L E

There are 45 geothermal power plants in the United States. Together, they produce enough electricity to supply the homes of more than 3.5 million people and save 2 million barrels of oil per year. If all the known geothermal resources in the country were developed, they could supply 27 times the total energy used in the United States every year.

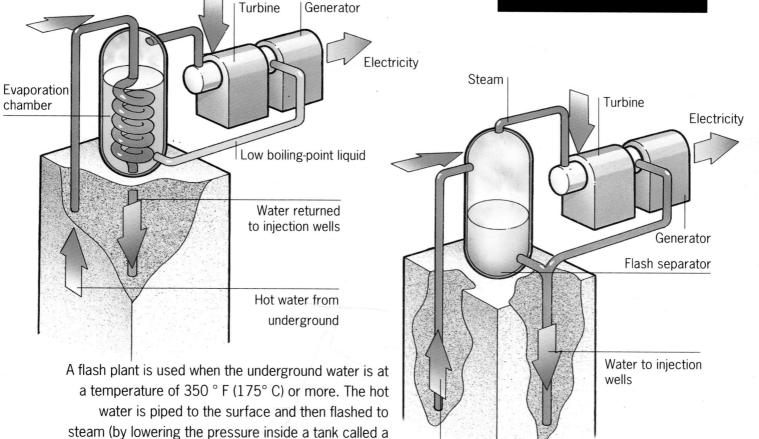

Turbine | Generator

Electricity

Evaporation chamber

Low boiling-point liquid

Water returned to injection wells

Hot water from underground

A flash plant is used when the underground water is at a temperature of 350 ° F (175° C) or more. The hot water is piped to the surface and then flashed to steam (by lowering the pressure inside a tank called a flash separator) to turn a turbine. The steam is then condensed into water and reinjected underground.

Steam

Turbine

Electricity

Generator

Flash separator

Water to injection wells

Hot water from underground

The Geysers Geothermal Power Plant

Geothermal power was produced in the United States for the first time in 1960 at The Geysers, an active volcanic area north of San Francisco, California. By 1967, The Geysers was producing electricity commercially.

Steam gushes from the The Geysers geothermal power plant in California's Sonoma and Lake counties. The power plant supplies electricity to people living in northern and central California.

Within six years, its output had grown from 54 megawatts to 412 megawatts, and it had become the world's largest geothermal plant. By 1986, the plant's output had reached 1,100 megawatts, enough to supply more than a million people. In 1994, it supplied 4.5 million megawatt-hours of electricity. At full power, it can produce nearly all the electricity needed by San Francisco.

Dry steam wells

The Geysers is fortunate in being able to use dry steam straight from the ground to drive its turbines and generators. The steam comes from 246 production wells through 55 mi. (88 km) of steam pipes.

After it has passed through the power plant, the steam is reinjected into the ground through 14 injection wells. The deepest of The Geysers' wells goes down nearly 2.5 mi. (4 km), and their average depth is 1.5 mi. (2.4 km).

At The Geysers site, steam constantly vents to the atmosphere from underground reservoirs through openings in the ground called fumaroles.

Above: To locate hot rocks underground, drill holes are made and probes sent down to measure the temperature.

Modern biofuels

Biomass can be converted into liquid fuels or gas fuels. As garbage in landfill sites break down naturally, they give off a gas that is a mixture of methane and carbon dioxide. This gas can be collected and used as a fuel.

A typical landfill site begins to produce gas after about three years and goes on doing so for up to 15 years. Better still, organic waste can be separated from general trash and broken down more efficiently in tanks called anaerobic digesters.

Wood-powered jet engines

Wood can be converted into gases that will power a gas-turbine (jet) engine. If wood is heated with a very limited oxygen supply, it gives off hydrogen, methane, ethene, carbon monoxide, and carbon dioxide. These gases are treated chemically to purify them, after which they are burned inside a gas-turbine engine.

The engine's hot exhaust gases are then passed through a boiler, which heats water. This produces enough steam to drive a turbine. The jet engine's turbine and the steam turbine both drive electricity generators.

A car in Brazil is filled with alcohol fuel—gasohol—made from biomass. Gasohol is a mixture of ordinary gasoline, made from crude oil, and ethanol produced from cassava and sugarcane. With little oxygen present, sugars extracted from these plants are broken down by yeast to yield energy.

A type of alcohol fuel called methyl alcohol, or methanol, can be made from wood. The wood is heated so that it gives off gases in a tank called a gasifier. The gases then go through a series of chemical reactions at high temperatures. The gas, or vapor, produced is compressed and distilled into a liquid—methanol.

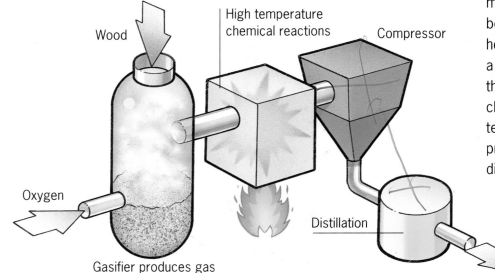

Wood

High temperature chemical reactions

Compressor

Oxygen

Distillation

Gasifier produces gas

Methanol

Some plants, such as these sunflowers in Spain, contain oils that can be extracted and used as a fuel to power machines or generate electricity. Each plant contains a tiny amount of oil, but a large crop of thousands of plants yields useful quantities of oil for fuel production.

THE FUTURE OF GEOTHERMAL AND BIO-ENERGY

The future of geothermal power

In Africa, several countries that are currently dependent on expensive foreign fuel and energy—Kenya and Djibouti for example—would like to develop their own geothermal power to reduce the cost of importing energy. Industrial countries in the northern hemisphere continue to develop their geothermal resources further. By the end of the twenty-first century, the United States could be getting almost one-third of its electricity from geothermal power plants.

Geothermal power plants, like this one at Ohaaki in New Zealand, may be common features in the future.

42

Geyser power

Just as a car's gasoline engine produces continuous power from a series of separate explosions, it may be possible to produce a continuous power output from a geyser or a number of geysers. Artificial geysers might be created for this purpose. By injecting water into holes drilled into hot rocks, artificial geysers could be used as steam generators for driving turbines.

People bathe in hot mud pools on Vulcano, one of the Aeolian Islands off Sicily, Italy. Doctors may increasingly recommend that sufferers of muscular pain bathe in hot pools, since this relieves their symptoms.

The future of bio-energy

Alcohol fuels made from plants will become increasingly important, especially to poorer countries that cannot afford to import all the coal and oil they need. While geothermal power can reduce the cost of a country's energy imports, biofuels produce something that can be sold to other countries.

Kenya is developing its sugar industry for the production of fuel oil. Miscanthus, a giant fast-growing grass from the Far East, could be the raw material for the biofuel of the future. It can be converted into more energy per acre than most other plants. Future filling stations might have gasohol, methanol, or biogas pumps alongside the traditional gasoline pumps.

Bio-energy from the sea

Most of the earth is covered by water. Seaweed flourishes in the fertile shallows around the continents. It represents a rich source of energy. In the future, seaweed may be farmed as a crop and converted into fuel and other useful chemicals.

In one experiment, kelp (brown seaweed) was grown on a vast web of cables in the sea off California. It was harvested and processed to make methane gas, animal feed, and a variety of chemicals.

In the next century, more of our energy needs will be met by plants. Local biogas digesters will convert farm and household waste into fuel gas for heating and making electricity. Crops will also be grown to be burned directly or converted into liquid fuels such as methanol.

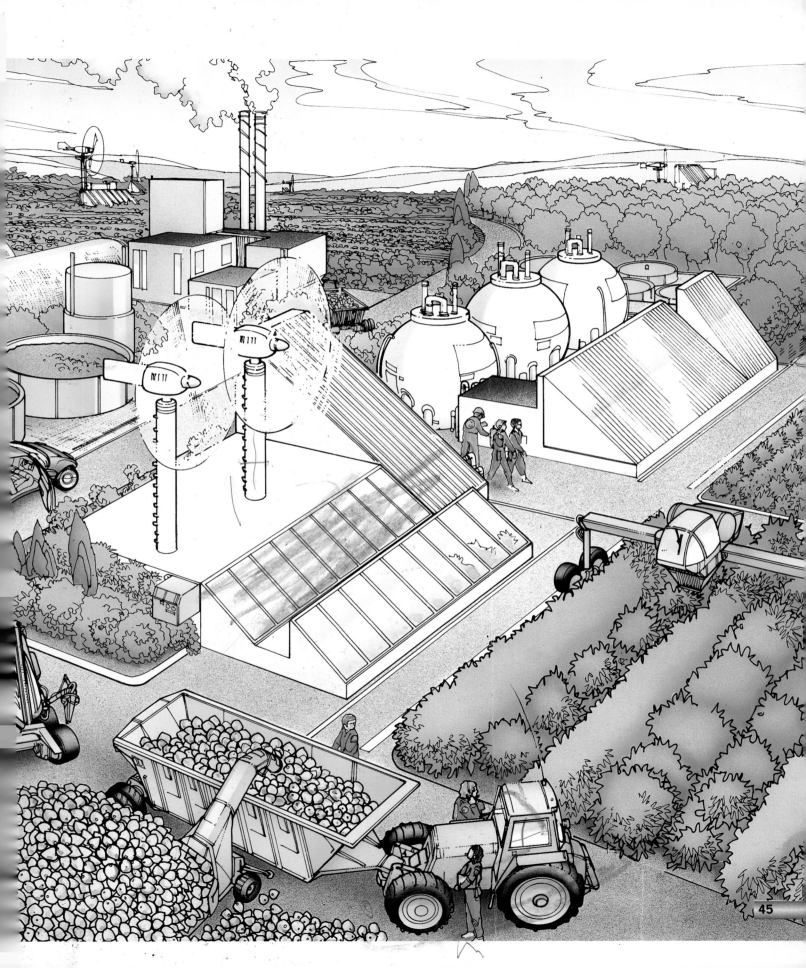

GLOSSARY

Air pollution
Unwanted or harmful gases and particles in the air.

Alcohol
A colorless liquid that burns easily and so can be used as a fuel.

Bacteria
Microscopic, single-celled organisms.

Barrel
(of petroluem) forty-two gallons (159 l)

Biofuel
A fuel made from plants or animal waste.

Biogas
A mixture of methane and carbon dioxide made by letting bacteria digest farm waste.

Carbon
A chemical element found in all living things.

Crude oil
Natural oil as it comes out of the ground, before it is refined.

Distillation
The process of purifying a substance by heating it so that it evaporates and then cooling it so that it changes back into a liquid.

Energy
The ability to do work.

Environment
The natural world around us.

Ethanol
A type of alcohol, also called ethyl alcohol.

Fermentation
A breakdown of biomass material by the action of microscopic organisms such as yeasts. Beer and bread are made by fermentation.

Fossil fuel
A fuel such as coal, oil, or natural gas formed from the remains of microscopic plants and animals that lived millions of years ago.

Fuel
A material that is burned to release the energy stored inside it so that it can be used for heating, generating electricity, or driving machinery.

Gasohol
A mixture of alcohol (usually made from plants) and gasoline. Gasohol can be used as fuel for motor vehicles.

Generator
A machine designed to change the movement energy of a spinning shaft into electricity.

Global warming
Warming of the earth's atmosphere as gases such as carbon dioxide—mostly produced by burning fossil fuels—trapping the sun's heat.

Greenhouse gas
Any gas, such as carbon dioxide, that traps the sun's heat and thus contributes to global warming.

Joule
A unit of energy.

Megawatt
A measure of electrical power equal to 1 million watts.

Methane
A flammable gas found in natural gas, also called marsh gas because it bubbles up through stagnant marshes.

Methanol
A type of alcohol, also called methyl alcohol.

Natural gas
Gas found in nature, usually in deep underground pockets, often with crude oil.

Power plant
A building where energy from a fuel is used to make electricity.

Turbine
Angled blades fitted to a shaft that is free to rotate. A moving gas or liquid flowing through a turbine presses against the blades and makes the turbine rotate.

Vapor
The gas produced when a liquid boils and evaporates.

Watt
A unit of power equal to the work done at the rate of 1 joule of energy in 1 second.

Watt-hour
A unit of energy equal to the power of one watt operating for 1 hour. A 100-watt light bulb burning for 1 hour consumes 100 watt-hours of energy.

Yeast
A microscopic single-cell fungus used in fermentation.

Books to read

Brown, Warren and Russell E. Train. *Alternative Sources of Energy* (Earth at Risk). New York; Chelsea House, 1993.

Challoner, Jack. *Energy* (Eyewitness Science). New York: DK Publishing Inc., 1993.

Chandler, Gary and Kevin Graham. *Alternative Energy Sources* (Making a Better World). New York: 21st Century Books, 1996.

Edmond, Alex. *The Greenhouse Effect* (A Closer Look At). Ridgefield, CT: Copper Beech Books, 1997.

Johnson, L. Rebecca. *The Greenhouse Effect* (Revised Edition). Minneapolis, MN: Lerner Publications, 1994.

Parker, Steve. *Energy* (Science Works). Milwaukee, WI: Gareth Stevens, 1997.

Power and energy consumption

Power is the measurement of how quickly energy is used. It is measured in joules per second, or watts. An electric iron might need 1,000 watts to work, but a portable radio might need only 10 watts. The energy needed to keep the radio going for one hour would run the iron for only six minutes, because the iron uses up energy ten times faster than the radio. The diagram to the right compares the power ratings of household electrical goods and of homes and power stations.

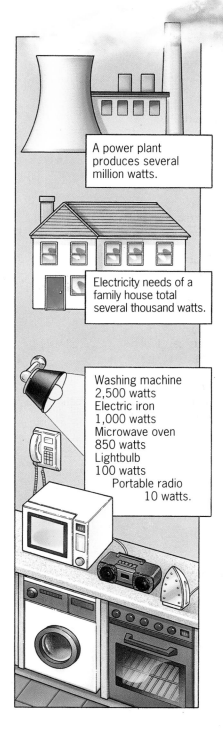

A power plant produces several million watts.

Electricity needs of a family house total several thousand watts.

Washing machine 2,500 watts
Electric iron 1,000 watts
Microwave oven 850 watts
Lightbulb 100 watts
Portable radio 10 watts.

INDEX